Mr. Fudd woke up and looked at the clock. "Good grief! I've overslept!" he said in disbelief.

"If I am late for work, my boss will be unhappy!"

Mr. Fudd threw on his clothes uncommonly fast. He did not see that his laces were still undone.

Then he rushed off to preheat a pan of water and cook some oatmeal.

Mr. Fudd was in such a hurry that he forgot to turn on the stove. So when he scooped up a big bite of oatmeal, he said, "Yuck! I must have undercooked this."

Mr. Fudd ran to the bus stop. "Why are there so many children and parents here? This is quite irregular," he said to himself.

Mr. Fudd did not know he was at the incorrect bus stop. This was a preschool group going on a field trip!

The bus was filled with lively chatter.

"What impolite children," Mr. Fudd said. "I will try to overlook the noise, but I certainly dislike it!"

"Do you like my pet mouse? His name is Chuck. Mine is Sam," said Sam.

Mr. Fudd said, "I dislike mice. Chuck should be undercover, I think."

With that, the mouse jumped down. Mr. Fudd jumped up. And soon there was a tangle of laces!

The bus driver stopped the bus. She said, "It is unsafe for me to drive. You are being too disorderly! Please settle down."

Sam put Chuck away. Then Mr. Fudd could untangle his laces. But he found it impossible to stay awake. He was overcome with sleep.

Mr. Fudd's alarm went off.

"It was all a dream, an impossible dream!" said Mr. Fudd.

Mr. Fudd was so overjoyed that he put on unmatched socks. But Mr. Fudd did not care! He was uncommonly happy!

The End

Understanding the Story

Questions are to be read aloud by a teacher or parent.

1. Why did Mr. Fudd make so many mistakes?
2. How do you think Mr. Fudd felt on the bus?
3. Why did the bus driver stop the bus?
4. Why was Mr. Fudd so happy at the end?

Answers: 1. because he was in a hurry 2. Possible answer: annoyed, upset 3. because it was unsafe to drive 4. because the bad dream was over

© Saxon Publishers, Inc., and Lorna Simmons

All rights reserved. No part of the material protected by this copyright may be reproduced or utilized in any form or by any means, in whole or in part, without permission in writing from the copyright owner. Requests for permission should be mailed to: Copyright Permissions, Harcourt Achieve Inc., P.O. Box 27010, Austin, Texas 78755.

Published by Harcourt Achieve Inc.

Saxon is a trademark of Harcourt Achieve Inc.

Printed in the United States of America
ISBN: 1-59141-038-X

1 2 3 4 5 446 09 08 07 06 05

Phonetic Concepts Practiced

pre- (preheat)
under- (undercooked)
over- (overslept)
un- (undone)
dis- (disbelief)
in- (incorrect)
im- (impolite)
ir- (irregular)

ISBN 1-59141-038-X

Grade 2, Decodable Reader 24
First used in Lesson 127

29

Tiny Plants, Big Plants

written by Holly Melton
illustrated by Janet Skiles

SAXON
PUBLISHERS

THIS BOOK IS THE PROPERTY OF:

STATE_____	Book No. _____
PROVINCE_____	Enter information
COUNTY_____	in spaces
PARISH_____	to the left as
SCHOOL DISTRICT___	instructed
OTHER_____	

ISSUED TO	Year Used	CONDITION	
		ISSUED	RETURNED

PUPILS to whom this textbook is issued must not write on any page or mark any part of it in any way, consumable textbooks excepted.

1. Teachers should see that the pupil's name is clearly written in ink in the spaces above in every book issued.
2. The following terms should be used in recording the condition of the book: New; Good; Fair; Poor; Bad.